Teacher's Notes
Interdependence and Adaptation

Merryn Kent

Series editor | Sue Palmer

Contents

OXFORD
UNIVERSITY PRESS

OXFORD

UNIVERSITY PRESS

Great Clarendon Street, Oxford OX2 6DP

Oxford University Press is a department of the University of Oxford.
It furthers the University's objective of excellence in research, scholarship,
and education by publishing worldwide in

Oxford   New York

Auckland Cape Town dar es Salaam Hong Kong Karachi
Kuala Lumpur Madrid Melbourne Mexico City Nairobi
New Delhi Shanghai Taipei Toronto

With offices in

Argentina Austria Brazil Chile Czech Republic France Greece
Guatemala Hungary Italy Japan Poland Portugal Singapore
South Korea Switzerland Thiland Turkey Ukraine Vietnam

British Library Cataloguing in Publication Data

Data available

ISBN-13: 978-0-19-834875-7
ISBN-10: 0-19-834875-4

7 9 10 8 6

Ashford Colour Press Ltd, Gosport, Hants

Printed in the UK

What is Oxford Connections?

Oxford Connections is a set of 12 cross-curricular books and related teaching materials for 7 to 11 year olds. The books will help you teach literacy through a science, geography or history-based topic. Each book provides the material to cover one unit from the QCA Schemes of Work for the National Curriculum in England and Wales, and the non-fiction literacy objectives for one whole year of the National Literacy Strategy. (You can find a grid of where the QCA and NLS objectives are covered on p 48 of these notes and on the inside back cover of the pupils' books.) The books can be used to focus primarily on literacy or on science/geography/history.

Literacy

Pupils need different literacies. As well as traditional texts with different purposes and audiences, they also need to be able to understand and write material presented in different forms such as diagrams, bullet points, notes and Internet displays, particularly when working with non-fiction.

Oxford Connections supports the development of these different literacies. It focuses particularly on reading and writing non-fiction, and will help pupils use effectively the different non-fiction text types (report, explanation, instructions, recount, discussion, persuasion).

Using these books will help pupils to focus on the two main elements which make a text type what it is:

◆ The language features used (for example, present tense for instructions, and past tense for recounts, use of commands in instructions).
◆ The structure of the text (for example, chronological order, in the case of instructions or recounts).

The structure of a text can be represented as a diagram or framework, showing visually how the parts of the text fit together, which are the main points and how they are developed. (A very common example of this type of presentation is a timeline, which shows events which have happened in the past, as a continuum, the order of which cannot change.) In these notes, we refer to material presented in this diagrammatic way as *visual* (visual reports, visual explanations, etc.).

Pupils will learn to read and to present information visually (by using frameworks), thus developing good note-taking skills, and consolidating their understanding of how texts are structured. The visual texts in particular are accessible to pupils who need more support. Using frameworks to plan their own writing will also help improve all pupils' planning and drafting/editing skills.

In these notes, we have used icons to represent the different sorts of frameworks you can use, called *skeletons*. These are referred to in the *National Literacy Strategy Support Materials for Text Level Objectives* (DfES 0532/2001). They can be used as an aide-memoir to help pupils remember the structure of each text type. They appear on pp 6–47 to show you what text types are on the pupils' book pages.

Recount		Explanation	
Instructions		Persuasion	
Non-chronological report		Discussion	

Using *Interdependence and Adaptation* to teach literacy

There are step-by-step instructions to teach pupils how to read and write the different text types on pp 18–47 (a six-page section for each text type). They follow this model:

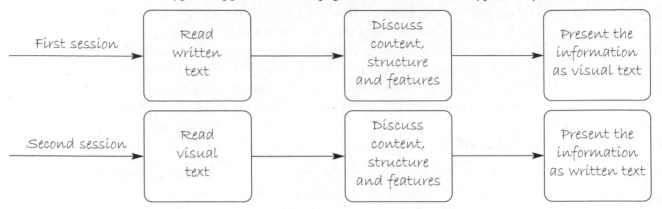

Each six-page section contains:

Two pages of step-by-step instructions taking you through the process described in the diagram above. They will help you analyse a written text, and then produce a visual version of that text with a group of pupils. You will then analyse a visual text, producing a written version.

A page describing the relevant text type.*

An example of the text type (an excerpt from *Interdependence and Adaptation*) for you to read and analyse with pupils.*

The same example with language features highlighted for your reference.*

A visual version of the written text for your reference.*

these page can be photocopied as handouts, a poster or an OHT

There are page-by-page notes on how to use the material to cover other aspects of literacy on pp 6–17. These page-by-page notes also show how to use the material in the pupils' book for the particular subject covered, e.g. science.

Speaking and listening, and drama

The discussion which is inherent in this method of learning should improve pupils' speaking and listening skills. As well as helping pupils to organize and structure their ideas before writing, visual texts should prompt pupils to use the relevant language features orally, as well as in writing. Additional speaking and listening, and drama activities such as those below, can be used to further reinforce the pupils' learning.

Retelling – events can be retold by an individual or by groups taking a section from a visual recount.

Role-play – using the visuals created by the whole class to ask/answer questions in role, as the person in the recount or taking one side of the argument etc.

Mini plays – retelling an event or following an explanation visual to show how something works. Pupils could be the different parts of whatever is being explained.

Puppet plays – retelling an event or following an explanation visual.

Freeze-frame – pupils in groups could show sections from a recount visual or report visual. They could show different aspects of a discussion.

TV/radio reports – demonstrating knowledge using a visual report as a TV/radio report. In a TV report images could be used either pictorially or by using freeze-framing.

TV demonstrations – following an instruction visual or explanation visual to demonstrate making something or explaining how something works.

TV/radio interviews – retelling events in recounts or using report visuals while interviewing another pupil/pupils in role.

TV/radio adverts – using a persuasive visual to make adverts.

Illustrated talks – using the visual as a prompt.

Hot seat – answering questions in role – either as a persuasion, report or recount.

Debates – using discussion visuals to have debates between individuals or groups.

Using *Interdependence and Adaptation* to teach Science

Interdependence and Adaptation contains all the material you need to cover this topic, and to achieve the objectives of the *QCA Scheme of Work for the National Curriculum Science Unit 6a* (recommended for Year 6 pupils). There are page-by-page notes on how to use the material for science on pp 6–17. You can find a grid showing where the QCA objectives are covered on p 48 of these notes, and on the inside back cover of *Interdependence and Adaptation*.

Which year group should I use *Interdependence and Adaptation* with?

Interdependence and Adaptation has been written for Year 6 pupils (10–11 year olds). However, if your school places the topic in another year group, the science material contained in *Interdependence and Adaptation* will still be suitable for use with other age groups. Although all of the non-fiction literacy objectives for Year 6 are covered, many of the objectives for other year groups are also supported. Most of the six non-fiction text types are covered in it, and language features for Years 3, 4, and 5 are highlighted in the relevant sections.

NB Throughout this introduction the term *Year 6* has been used to mean 10–11 year olds. The references in the grid on p 48 are to the *National Literacy Strategy* and to the *QCA Scheme of Work for the National Curriculum*. However, *Interdependence and Adaptation* is suitable for use with P7 in Scotland and in Northern Ireland, since it supports many elements of the *National Guidelines, 5–14* and *The Northern Ireland Curriculum*. The science content of *Interdependence and Adaptation* does not conflict in any way with either *National Guidelines, 5–14* or *The Northern Ireland Curriculum*.

SCOTLAND AND NORTHERN IRELAND

Science

Use these pages as advance organizers to provide pupils with an overview of the work to be carried out:

Concept map shows the main areas to be covered and links between them;

Contents page shows how information has been organized in the book.

◆ Use the quotation as an aid for using the contents page – ask pupils to find what is being mentioned.

◆ Return to these pages occasionally during teaching to help pupils see how their learning and understanding is building up.

◆ Use these pages as a revision aid, asking pupils to summarize what they know about each aspect.

◆ Use the concept map at the end of the topic to review all areas of the topic covered.

Literacy

Help pupils to recognize the similarities and differences between the concept map and the contents page.

◆ They contain the same information but are organized differently.

◆ The concept map provides an overview of the ideas in the book and shows how they are interlinked; the contents page provides a linear guide to the way these ideas are organized.

◆ The contents page gives page numbers for ease of reference.

Throughout your use of the book demonstrate how to use the contents page along with the index (see p 17 of these notes) – to access information.

Science

Key concepts

◆ The Sun is the source of energy for almost all life on Earth.

◆ Plants use the Sun's energy directly.

◆ Many other forms of fuel, including food, are derived from plants.

Key vocabulary

◆ *energy, life process, organism, organic*

Suggested activities

◆ Use these pages to establish the idea that all living things need energy, and that all energy for life on land is derived from the Sun.

◆ Collect examples of fuels e.g coal, charcoal, food, plant material. Using these and other examples, ask pupils to trace the source of energy back to the Sun.

Literacy

page 4	page 5
visual report	visual report

Use these pages to discuss the relative merits of this visual report compared with written reports alone for conveying information.

◆ Study the spidergram and discuss how ideas are linked together. Note also how the diagram, illustrations and notes add to the clarity of the information. Ask how else this information could be presented.

◆ Use the section 'What is the Sun?' to practise creative writing based on the Egyptian legend. In groups, pairs, or individually, pupils research and write their own 'legend' to explain the movement of the Sun. [Link to Fiction and History]

Science

Key concepts

◆ All living things need food.
◆ Plants and animals need energy.
◆ Plants and animals obtain food in different ways.
◆ Ultimately, plants and animals derive their energy from the Sun.

Key vocabulary

◆ *carbon dioxide, photosynthesis, chloroplast, producers, consumers*

Suggested activities

◆ Use pages to review what pupils remember about the conditions necessary for plant growth. If possible, show a healthy green plant and ask questions about it. Identify that water, carbon dioxide from the air, and sunlight are needed to make food. Warmth, healthy stems, leaves, and roots are needed for the plant to grow well.
◆ Introduce the topic of producers and consumers (see notes to pp 26–27 in the pupils' book).

Literacy

page 6	page 7
O→☼→♡	O→☼→♡
visual explanation	visual explanation

These pages are used as a featured example to teach pupils how to read and write **explanation** text (see pp 30–35 of these notes).

◆ In pairs, or groups, pupils discuss whether humans should be called producers or consumers.

Science

Key concepts

◆ Plants need light to grow well.
◆ Leaves are green because they contain chlorophyll.
◆ Healthy leaves that are not green still contain chlorophyll in their chloroplasts.
◆ Plants use sunlight to make their own food.

Key vocabulary

◆ *chlorophyll, chloroplasts, photosynthesis, starch*

Suggested activities

◆ Carry out the two experiments suggested.
◆ Ask the pupils to predict what will happen to the plant left in the dark if it is returned to the light and looked after. (Note – do not leave the plant in the dark for too long or it will not recover.)
◆ Observe and record results in a table. (Link to ICT)

Literacy

page 8	page 9
O→O→O→	O→O→O→
visual instructions	written instructions

Ask pupils to use the step-by-step illustrations on p 8 as the basis for a set of notes which they can present as a visual instructions skeleton. They then use the skeleton notes to produce a set of numbered written instructions describing how to conduct an experiment entitled e.g. *Do plants need light to grow?*

Use p 9 to remind the class of the language features (*e.g. imperative, present tense; numbered steps or time order; enough detail to follow directions; clear, concise language*) and organization (*e.g. what is to be achieved; list of items required; sequential steps*) of written instructions.

◆ To secure their understanding of the text type, pupils could write skeleton instruction notes for a simple recipe, presented as a flowchart.

Pages 10–11

Science

Key concepts
◆ Plants use water, carbon dioxide and light energy from the Sun to make starch.
◆ Plants produce oxygen when they photosynthesize.

Key vocabulary
◆ *photosynthesis, oxygen, carbon dioxide, interdependent*

Suggested activities
◆ Ask pupils to design experiments which could show that plants produce oxygen but do not harm animals (e.g. *bubbles released from Canadian pondweed*).
◆ Analyse one or two suggestions in more depth. Discuss whether they would satisfy certain criteria e.g *fair test, repeatable, measurable, accurate.*

Literacy

page 10	page 11
┼┼┼→	┼┼┼→
written recount	written recount

A reduced version of the text on these pages is used as a featured example to teach pupils how to read and write **recount** text (see pp 18–23 of these notes).

Also use to discuss:

◆ the difference between biography and autobiography;
◆ the use of third and first person;
◆ the difference between fact and opinion;
◆ how some words (e.g. 'phlogiston') have fallen out of use;
◆ why Ingenhousz created the word 'phlogiston' for his substance (from *phlogizein* – a Latinized version of the Greek word meaning 'to burn'); look at the origins of other technical words.

Pages 12–13

Science

Key concepts
◆ Soil has many components including living organisms.
◆ Different plants like different types of soil.
◆ Soil can be replaced by gravel, water and nutrients to grow plants.
◆ Microorganisms in soil have useful functions.

Key vocabulary
◆ *microscope, mineral, organic, nutrient, decompose, hydroponics*

Suggested activities
◆ Collect plant labels to use for a discussion about the needs of plants, including soil types.
◆ Examine two contrasting soil samples under a microscope or hand lens (e.g. *sand and clay soil*). Observe and record differences in a table. (Link to ICT)
◆ Ask pupils to deduce what kind of plants would prefer each soil type.

Safety Note **Be aware of the need for health and safety precautions when pupils are handling soil.**

Literacy

page 12	page 13
written and visual report	written report

Use for class discussion about different ways of presenting information as used in this spread. Ask why the author has chosen each one (e.g. *the comparative report in the grid allows comparisons at a glance; written report used to convey complex ideas*).

Pupils could also use the skeleton notes about microorganisms to produce a written report about 'Microorganisms that live in the soil'.

The pupils use the grid to produce a written comparative report about plants from different soils after making skeleton notes (see p 29 of these notes for an example of a report skeleton). They should pay attention to similarities and differences using appropriate linking phrases e.g *but, however, on the other hand.*

Science

Key concepts
◆ Plants take in nutrients through their roots.
◆ Roots anchor the plant.
◆ Some plants have adapted to their habitat by developing specialized roots.

Key vocabulary
◆ *anchoring, tap root, fibrous root, root hairs, nitrogen fixing nodules, tubers*

Suggested activities
◆ Use these pages to explain the function of roots.
◆ Also use to develop the pupils' understanding of *adaptation*, by studying the description of different root types.
◆ In the classroom, look at some plants that have a distinct tap root e.g. *carrot, parsnip, dandelion*.
◆ Let the pupils make an observational, still-life drawing of plants, showing their roots. Discuss and compare these pictures and their function with the diagram on p 14. (Links to QCA Art and Design Unit 5a)

Literacy

page 14	page 15
written and visual report	visual report/ reference

The visual report on page 15 can be used as a featured example to teach pupils how to read and write **reference** text (see pp 42–47 of these notes).

◆ Use written information and diagrams on p 14 to make skeleton **report** notes (see p 29 of these notes for an example of a report skeleton).
◆ Pupils use these notes to explain orally, with the aid of diagrams, what a root is and the various functions of a root.

Science

Key concepts
◆ Plants need certain nutrients to grow properly.
◆ Plants usually obtain these nutrients from soil.
◆ Specific deficiencies lead to specific symptoms.

Key vocabulary
◆ *food supplements, minerals, vitamins, nutrients, fertilizers*

Suggested activities
◆ Using information extracted from the text, pupils draw and label a healthy plant, and plants showing different deficiencies. This could be done in groups with each pupil in the group allocated a 'symptom'.
◆ Pupils could use reference books, health information or the Internet to find out which vitamins and minerals are needed to keep humans healthy. They could draw up a table to show the symptoms of nutrient deficiency. (Link to ICT)

Literacy

page 16	page 17
written report	written report

These pages are used as a featured example to teach pupils how to read and write **report** text (see pp 24–29 of these notes).

◆ In pairs or groups, pupils could role-play in a Speaking and Listening session, holding a discussion between gardeners and a plant 'doctor'. 'Gardeners' report problems with their plants and receive advice on treatment. (Link to Drama)
◆ Point out to pupils the visual report on p 17 in the form of a chart showing signs of nutrient deficiencies. They could use the information to make more detailed skeleton notes which could then be turned into a full written report.

Science

Key concepts

◆ In the past, farmers have modified crops by cross-breeding plants.

◆ Now, genes can be altered directly in a laboratory.

◆ Some see GM (genetic modification) as efficient and useful.

◆ Others see GM as a potential risk to health and the environment.

Key vocabulary

◆ *genetic modification (GM), gene, laboratory, cross-breeding, pesticide, insecticide, herbicide*

Suggested activities

Use this article from a journal to consider carefully the pros and cons of genetic modification. Hold a class vote. Do some more research (eg. on the Internet, from newspapers), have a class debate and hold the vote again. Has scientific evidence caused anybody to change their opinion? Encourage logical debate, and argument reinforced by evidence.

Literacy

page 18	page 19
✳✳|✳ ✳✳✳ ✳|✳	✳✳|✳ ✳✳✳ ✳✳|✳
written discussion	written discussion

These pages are used as a featured example to teach pupils how to read and write **discussion** text (see pp 36–41 of these notes).

◆ Draw pupils' attention to the suffix *–cide* at the end of *herbicide, insecticide* and *pesticide*. Investigate its meaning and find other words with the same suffix.

Safety Note **Some words with the –cide suffix have sinister or distressing connections, so be aware of any need for sensitivity within the classroom.**

Science

Key concepts

◆ Fertilizers contain nutrients needed by plants to remain healthy.

◆ Fertilizers have both advantages and disadvantages, especially when used on a large scale.

Key vocabulary

◆ *organic fertilizer, chemical fertilizer, nutrient, pollute*

Suggested activities

Grow two separate crops of one type of plant. (Tomatoes are good if it is the right time of year or try other plants.) Use an appropriate fertilizer or plant food on one crop, or group of plants, and *no* fertilizer on the other. Otherwise, care for them in the same way. Is there a difference in the growth rate, health or yield of the plants? Make and record observations. (See table on p 9 of the pupils' book – Link to ICT)

◆ What happens if a plant is given too much fertilizer?

Safety Note **For health and safety reasons pupils should not handle fertilizer themselves.**

Literacy

page 20	page 21
⬡	✳✳|✳ ✳✳✳ ✳✳|✳
written report	visual discussion

The visual discussion skeleton notes on p 21 are used as a featured example to teach pupils how to write **discussion** text (see pp 42–47 of these notes).

◆ In a Speaking and Listening activity, pupils could hold a debate on the advantages and disadvantages of fertilizers. Working in pairs, pupils could prepare short speeches to contribute, based on the skeleton notes on p 21 – one of the pair describing one advantage, while the other puts forward the opposing argument to that aspect of the subject.

Science

Key concepts
◆ Fertilizers increase crop yield.
◆ Misuse can lead to pollution.

Key vocabulary
◆ *fertilizer, run-off, nutrients, algae, toxins, nitrogen, oxygen*

Suggested activities
As a class, have a look at some packaging from fertilizers and plant food. Discuss why different plants require different types of fertilizer or food (refer back to pp 16–17 of the pupils' book if hints are necessary). Reintroduce the concept of *adaptation*.

Safety Note **For health and safety reasons, pupils should not handle fertilizer themselves.**

Literacy

page 22	page 23
✳≶ ✳≶ ✳≶	✳≶ ✳≶ ✳≶
written persuasion	visual persuasion

Read p 22 with the pupils. As a class, consider:
◆ What is the aim of the article?
◆ What language features indicate this? e.g. appealing to the emotions (the area may take years to recover); no mention of the benefits of fertilizers; use of persuasive examples.
◆ What is the impact of the illustration?

Now look at the advertisement on p 23 and analyse it in the same way considering what the advert wants to persuade you to do and how it does this.

As a class, make skeleton notes of your findings by making a series of bullet points about each advantage, elaborated if necessary giving details about each point. Pupils could then produce a piece of written persuasion text (e.g. a review in a magazine), recommending the use of this fertilizer.

◆ In pairs, pupils use their skeleton notes to make two posters, one urging people to cut down on the use of fertilizers; the other advertising the benefits. (Links to Art and DT)

Science

Key concepts
◆ Different animals have adapted to find and digest different foods.
◆ Three main types of feeder are herbivores, carnivores, and omnivores.
◆ Two other types of feeder are scavengers and detritivores.

Key vocabulary
◆ *herbivore, carnivore, omnivore, (and related vocabulary), scavenger, detritivore*

Suggested activities
◆ As a class, discuss and answer the question posed about which category humans fit into.
◆ Discuss the difference between a herbivore and a vegetarian/vegan.
◆ In groups or pairs, pupils choose some other animals, and decide what category of feeder they are.
◆ Extra research could be carried out if necessary using reference books or the Internet with the groups/pairs making a report to the whole class in a Speaking and Listening activity The findings could be presented as a comparative report in a chart or table (Link to ICT).
◆ Use this information to introduce the topic of food chains (see next spread pp 26–27).

Literacy

page 24	page 25
visual report	visual report

These pages are used as a featured example to teach pupils how to write **report** text (see pp 24–29 of these notes).

Science

Key concepts
- Feeding relationships can be represented by food chains and webs.
- The Sun's energy is used by plants and passed up the food chain.
- Some energy is lost as it passes up the chain.

Key vocabulary
- *producer, consumer, feeding relationship, food chain, food web, predator, prey*

Suggested activities
- Review the information on pp 4–5 and 6–7 of the pupils' book.
- Groups of pupils make 'food chain' cards. Each card group should contain cards showing a producer and two or more consumers. Pass the cards among groups to sort them into chains.
- As a circle-time activity, discuss Joseph Banks' method of collecting specimens (shooting them!) Widen the discussion. Include the role of zoos, conservation groups and animal welfare organizations, and our responsibility towards animals. (Link to Citizenship)

Literacy

page 26	page 27
๐→✿→๑	๐→✿→๑
written and visual explanation	written and visual explanation

- These pages are used as a featured example to teach pupils how to read and write **explanation** text (see pp 30–35 of these notes).
- Pupils make skeleton notes, based on *one* of the food chain or food web diagrams. They use their notes to turn these visual explanations into a written explanation.
- Using Banks' historical journal recount, pupils make skeleton notes, then rewrite them in the style of a modern journal entry.
- Use the extract from Joseph Banks' journal to find variations in spelling from accepted modern spelling; study features of the writing which indicate it was written some time in the past e.g. *afforded* a far larger variety, we saw *but few*.

Science

Key concepts
- Plants and animals have adapted to exist together.
- Without plants, the animals in the world would be different.
- Without animals, the plants in the world would be different.
- There is a cycle of growth, death and decay which recycles nutrients.

Key vocabulary
- *cycle, evolved, co-existence, recycle*

Suggested activities
Divide class into groups or pairs. Give each group the name of a plant and animal with which they are familiar or can research. Using reference books or the Internet, each group finds out whether their plant has any adaptations as a result of interdependence with animals (e.g. *flowers for pollination, spikes to deter browsers*). Similarly for the animal. The groups or pairs come together to discuss and compare findings. This work could be limited to particular habitat(s) known to the pupils. The results could be displayed in a comparative report on a chart or table (Link to ICT).

Literacy

page 28	page 29
๐→✿→๑	๐→✿→๑
written explanation	written and visual explanation

- This report/explanation text could be used as a basis for a discussion about a world without plants/animals in a Speaking and Listening activity. On the whiteboard, make notes as a class to summarize the main points. Pupils could then make their own notes and produce some imaginative prose/poetry describing such a world.
- Pupils can use the visual explanation on p 29 to produce a written explanation of the ecological cycle. First, adapt the information on the diagram to make skeleton notes.

Pages 30–31

Science

Key concepts
◆ There are so many different species on Earth, it is sometimes hard to tell them apart.
◆ 'Identification keys' help scientists to sort organisms into groups.

Key vocabulary
◆ *species, identification key, spider (tree) key, numbered key*

Suggested activities
◆ Name a group of objects (e.g. a group of red objects, a group of rectangular objects) and ask the pupils what they have in common. Point out that this is how scientists group organisms – they have something in common. Model using the keys, then do an example together.
◆ Remind pupils of a habitat they have visited before (they should have studied a local habitat in Year 4). If possible revisit it. Use a key to identify some plants and animals found here and display the results. This could be an ICT-based activity (Link to ICT QCA Unit 4c).

Literacy

page 30	page 31
○→▷→○	○→▷→○
written explanation	visual explanation

In pairs, pupils write explanation skeleton notes about how to use a key for the above activity. Once pupils are confident using keys, they can practise instruction writing based on the written explanation about keys given in the text. They could use the notes to write instructions for a Year 4 class, or to demonstrate the procedure to a Year 4 class in a 'show and tell' session.

Point out the similarities and differences between explanation and instruction texts, including their 'skeletons', e.g. *chronological order of the procedure to be followed; numbering system; use of imperative; present tense.*

Pages 32–33

Science

Key concepts
◆ Animals and plants are adapted to their environment.
◆ An innate feature that provides an advantage tends to be passed to offspring.
◆ A characteristic acquired during a lifetime cannot be passed to offspring.

Key vocabulary
◆ *adapted, adaptation, inherit*

Suggested activities
◆ Carefully study the animals described in the visual report to see how they are adapted to their habitat and lifestyle.
◆ After class discussion, ask pupils to choose an environment. Let them draw/paint an animal of their own invention which is adapted to the environment they have chosen (Link to Art). They then write a brief description of the animal and the features that make it especially suited to its environment and lifestyle.
◆ Alternatively, they could invent an 'extinct' animal which has features which are poorly adapted to its environment, e.g. a cat which walks slowly after prey, or makes a lot of noise while hunting.

Literacy

page 32	page 33
✲○✲	○→▷→○
visual report	visual explanation

◆ In pairs, use the visual report on p 32 of the pupils' book to make skeleton notes (see p 29 of these notes for an example of a report skeleton) of the three animals mentioned.
◆ As an independent exercise, pupils turn the skeleton notes into a written report on the animals and how they are adapted to their environment and lifestyle.
◆ Using p 33, pupils study the two visual explanations and use the notes to produce a piece of written explanation about each theory. Some pupils might then be able to write a concluding paragraph explaining why the Darwinian explanation has become accepted as the correct theory.

Science

Key concepts

◆ Darwin proposed that species evolve through a process known as natural selection.

◆ Variations that give an advantage are passed on to offspring.

◆ In time these changes lead to the forming of new species.

Key vocabulary

◆ *evolve, evolution, natural variation, natural selection, adaptation, species, survival, extinct*

Suggested activities

◆ Point out that other scientists had theories about evolution. (See Lamarck's theory on pp 32–33 in the pupils' book.) Charles Darwin was radical in proposing that evolution came about through random, natural selection. (Some links to History QCA Units 11–12)

◆ Have a class discussion about animals that once thrived but are now extinct e.g. *dodo, dinosaurs, passenger pigeon.* Animals are adapted to their environment but evolution happens slowly – if a change occurs too quickly, animals and plants are threatened.

Literacy

page 34	page 35
written report	written report

◆ Using skimming techniques, quickly read through the written report text on pp 34–35.

◆ As a class, discuss the text.

◆ Using p 34, pick out key words and phrases in pairs (see p 47 of these notes).

◆ As an independent activity, pupils summarize each section orally, or in writing, from their notes, using one or two sentences.

◆ As a class, discuss any problems which arose (e.g. *difficult vocabulary; complex ideas*).

◆ Using p 35, discuss the style of language used by Darwin in these extracts, such as words and expressions rarely used today – e.g. *let us suppose, fleetest, hard-pressed, supplant.*

Science

Key concepts

◆ Darwin's observations while on the *Beagle* helped him to develop the theory of natural selection.

◆ Publication of his work led to opposition from the Church.

◆ Darwin's theory of evolution became accepted in scientific circles.

Key vocabulary

◆ *naturalist, theory of evolution, human evolution*

Suggested actvities

◆ Study the map of the route of HMS *Beagle.*

◆ Using an atlas, reference books or the Internet, find out where the Galapagos Islands are, what the climate and environment is, and what kind of animals live there.

◆ Use an atlas to discuss why the animals of the Galapagos developed into different species. (Links to Geography QCA Units 18 and 24)

Literacy

Page 36	Page 37
visual recount	visual recount

These pages are used as a featured example of how to write biographical-**recount** text (see pp 18–23 of these notes).

◆ Working in groups, pupils write and perform a short play or a series of freeze-frames about the life of Darwin. (Link to Drama)

Science

Key concepts

◆ Different animals and plants live in different habitats.

◆ Within a habitat each animal occupies a specific ecological niche.

◆ Similar animals can live in the same habitat if they occupy a different niche.

◆ Animals can occupy the same niche and live in different habitats.

◆ Animals occupying the same niche in the same habitat compete for resources.

Key vocabulary

◆ *habitat, environment, ecological niche*

Suggested activities

◆ Visit a habitat unfamiliar to the pupils, or study one using secondary sources (e.g. books, videos, Internet, CD-ROMs). Observe/study animals and plants in that habitat. Consider these questions:
  - How are these animals and plants adapted to their environment?
  - How numerous are they within the habitat?
  - What is their ecological niche? Are they competing with other species?

Make a class display of the habitat based on answers.

◆ See also – dodo on p 35 of the pupils' book.

Literacy

page 38	page 39
written report	written report

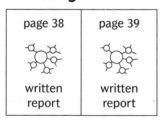

Use pp 38–39 to produce a comparative report by making notes to go on a characteristic grid for each section. Discuss what headings need to go into the grid, and how to record information. Demonstrate on the whiteboard or an OHT, showing how to complete a grid (e.g. about crossbills and siskins); complete another (e.g. about gazelles and kangaroos) as a piece of shared note-making; the pupils make notes to complete the third as an independent exercise.

◆ Pupils use the grids to answer comprehension questions orally.

Science

Key concepts

◆ The panda is not well adapted to its ecological niche.

◆ Efforts to protect the panda may do more harm than good.

Key vocabulary

◆ *deforestation, nature reserve*

Suggested activities

◆ Look at some bamboo. Discuss briefly why this is not an efficient food for an animal that was originally a carnivore. (See p 24 for an example of a typical mammalian herbivore and carnivore.)

◆ Find something that weighs about 18 kg so pupils realize how much this is. How many kilograms of bamboo would a panda have to eat in a week/a month/a year in order to survive? (Link to NNS Maths)

Literacy

page 40	page 41
written report	written report

The information in these two written report texts could be used as the basis for a piece of persuasive writing.

◆ Tell the pupils that Tom has decided to write to his tour company, to persuade them not to visit the Woolong Nature Reserve, or to use its hotels.

◆ Use pp 40–41 to make skeleton persuasion notes by writing a list of bullet points, expanded if necessary.

◆ Remind pupils that they are using two different sources. Compare the type of language used in each (e.g. *formal/informal; general/personal, third person/first person*).

◆ Discuss the style and language features needed for Tom's new letter e.g. *main points; technical words and details to support argument; strong, emotive language etc.* Encourage the pupils to extract only relevant information.

Science

Key concepts
- Animals are adapted to their environment and lifestyle.
- Animals in extreme conditions such as the tundra may adapt to the environment in similar ways.
- Animals and plants in a habitat are interdependent.

Key vocabulary
- *permafrost, adaptation, insulation, camouflage*

Suggested activities

- Use this and the following spread to summarize key concepts and vocabulary within the topic of Interdependence and Adaptation.
- In pairs or groups, discuss why animals in the tundra share certain adaptations (e.g. extreme temperatures; need to conserve energy because of food shortages and cold; need for white camouflage in winter months). Bring ideas together in a class discussion.
- Use reference books or the Internet to find out about producers (plants) in the tundra (Link to ICT). Construct food chains or webs that include these producers, and primary and secondary consumers (animals) mentioned in the text.
- Discuss ways in which the producers limit the number and type of consumers in the tundra.

Literacy

page 42	page 43
written report	visual report

Use these pages for report writing (see pp 24-29 of these notes). Once the Science activities suggested above have been completed, pupils could turn their findings into skeleton notes to produce a written report of why and how animals have adapted to life in the tundra.

Science

Key concepts
- Different habitats support different plants and animals.
- Unique species may develop on islands, such as Madagascar.
- Habitats rich in vegetation can support a wide variety of animals.

Key vocabulary
- *rainforest, desert, unique, habitat, environment, conservation, endangered*

Suggested activities
- Identify feeding types (carnivores, herbivores, omnivores, scavengers, detritivores).
- Find out about other habitats on the island (Links to ICT and QCA Geography Unit 24). Use findings to review the concept that different animals and plants are adapted to live in different habitats.

Literacy

page 44	page 45
written report	visual report

After completing the activities above, use these pages to reinforce writing skills. For example:

- a written/spoken **explanation** of food chains and/or food webs;
- a **report** on feeding types in the Madagascan rainforest;
- a **recount** in journalistic style describing a trip round the island;
- a piece of **persuasive** writing recommending wildlife-viewing holidays in Madagascar;
- a **discussion** text detailing arguments for and against tourism in Madagascar.

Pupil should write skeleton notes in pairs, then write their chosen activity independently. This work and that from the previous pages could form an oral class presentation (e.g. an assembly) or be written up as a class book to summarize what pupils have learned in the Interdependence and Adaptation unit of work.

Science

- Use the **glossary** to check understanding of terms used and as an aide-memoir.
- Use the **bibliography** for cross-checking information and for additional research.

Literacy

page 46	page 47
○ ↓ ○ ↓ ○	○ ↓ ○ ↓ ○
written reference	written reference

P 46 in the pupils' book is used as a featured example to show pupils how to read and write **reference** texts (see pp 42–47 of these notes).

Discuss the bibliography (including the word and its origins). Consider these questions:

- What do you notice about: a) its organization; b) its content?
- What would you use a bibliography for?
- Why is a bibliography included in many information texts but not usually in works of fiction?

Science

- Use the **index** to locate topics on individual pages.
- Use the index to revise work on replacement texts already covered, and for cross-referencing.

Literacy

page 48
○ ↓ ○ ↓ ○
written reference

Use this page in addition to the glossary as an example to show pupils how to read and write alphabetic **reference** text (see pp 42–47 of these notes).

Teaching pupils how to read and write biographical recount text

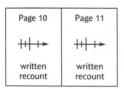

Reading a recount text

Read pp 10–11 of *Interdependence and Adaptation* with the pupils. You will need:

- the written recount on pp 10–11 (the reduced, text-only version on p 21 of these notes can be enlarged/photocopied/made into an OHT for annotation – NB some of the scientific detail has been omitted from the version on p 21, see pupils' book for full text);
- p 20 of these notes enlarged/photocopied/made into an OHT for annotation.

Audience and purpose

> SHARED
> READING
> ACTIVITY

Talk about how the intended audience and purpose affects language and layout.

Audience – pupils from another (or the same) class who may not know much about these events.

Purpose – to retell events in the life of Ingenhousz that actually happened.

Content and organization

> SHARED
> WRITING
> ACTIVITY

Demonstrate to pupils how the information in the recount text is organized by showing its content in a timeline (see p 23 of these notes). The timeline can be divided into sections. Each section can be turned into a paragraph in a written recount.

Language features and style

> SHARED
> READING
> ACTIVITY

Return to the text and talk about the way language has been used to achieve the effects the author intended (see annotated version on p 22 of these notes). Note useful features for later use in pupils' own writing e.g. use of words to do with time and time passing (*first, then, while);* consistent use of past tense. Add other examples provided by the pupils.

> INDEPENDENT
> PAIRED OR
> GROUP
> ACTIVITY

As a further exercise, divide the class into pairs or groups. Each group or pair has a section of the timeline to study and develop. In a whole-class session, the groups/pairs answer questions about 'their' period in the life of Ingenhousz. This could be done as a role-play session if desired. (Link to Drama)

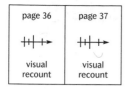

page 36	page 37
visual recount	visual recount

Writing a recount text

Use pp 36–37 of *Interdependence and Adaptation* as a basis for pupils' own written recount. You will need:

◆ the visual recount on pp 36–37;
◆ p 20 of these notes enlarged/photocopied/made into an OHT for annotation.

Content and organization

> SHARED READING ACTIVITY

Study the timeline with the pupils, using questions and answers to encourage them to become familiar with the layout of the timeline and the progress of Darwin's work. Point out that the timeline emphasizes one aspect of Darwin's life, showing how his work on HMS *Beagle* helped him to develop his theory of natural selection.

> PAIRED READING ACTIVITY

In pairs, pupils use the notes on the skeleton timeline to help them retell sections of the story to each other. The notes from each section will become a paragraph of written recount. They should therefore consider how to divide the timeline so that each section covers an important stage in Darwin's life – e.g. 1809–1827; 1831–1839; 1859–1871; 1882.

Language features and style

> SHARED READING ACTIVITY

Remind pupils of the language features and style of written recounts (see p 20 of these notes).

Audience and purpose

Discuss the audience for pupils' recounts (readers who know little about the life of Darwin) and purpose (to provide a factual, chronological account of significant events in Darwin's life).

Demonstrate writing a paragraph, for example:

Charles Darwin was born in Shrewsbury in 1809. At the age of twenty-six he went to University in Edinburgh to study medicine but, as he hated the sight of blood, Darwin did not do well. In December the following year, he swapped his medical vocation for that of the Church and attended Cambridge University hoping to become a clergyman. Any free time Darwin had was spent socializing – and collecting beetles!

> SHARED WRITING ACTIVITY

Scribe a paragraph with the pupils, making sure that the paragraph describes a significant event or period in Darwin's life (see sections suggested in paired reading activity above).

> INDEPENDENT WRITING ACTIVITY

Pupils work on remaining paragraphs independently, or in pairs, to create a complete piece of written biographical recount text about Darwin.

About recount text

Audience and purpose

Audience – someone who may not know much about the events.

Purpose – to retell events that actually happened.

> Sometimes you may know more about the age or interests of your reader

Content and organization

- **introductory paragraph** sets the scene, so the reader has all the basic facts needed to understand the recount
- **introduction** often also hints at the main event of the recount
- events written in **chronological order** – time order
- **closing statement** – sentence(s) or paragraph to bring the recount to an end

> Answer the questions who? what? when? where?

> Use your introductory sentence to help you write your conclusion. If the introduction is a question then answer it in your conclusion

> First this happened...then this happened... next...

Language features

- written in the **past tense** because these are specific events that only happened once
- focus on **specific people, places, dates** etc.
- may be written in the **first** or **third person**
- **words and devices** to show **time order**

> This usually means proper nouns, so remember the capital letters!

> Stick to one or the other – don't mix them up

> First..., next..., finally..., In 1950..., Some weeks later...

The basic skeleton for making notes is a timeline

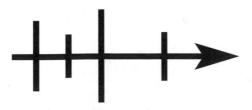

An example of a recount text

The Life of Jan Ingenhousz

Jan Ingenhousz was born in Breda, Holland in 1730. He was the son of a leather merchant. Jan was a clever boy and he ended up going to university to study medicine first in Holland, then in Paris and Edinburgh.

He worked as a doctor in his home town until he was 35, when he went to work in a hospital in London. Here he was one of the pioneers of inoculation against the deadly disease, smallpox. In 1768 his work was noticed by the king, George III, who sent him to Vienna to inoculate the Austrian Empress Maria Theresa. Ingenhousz stayed in Austria as court physician until 1779, when he returned to England.

While he was in Austria he studied other subjects, including electricity and plants. His work on plants was very detailed and he did hundreds of experiments to try and find out exactly how they worked.

Ingenhousz had met Priestley when he was working in London and knew all about Priestley's work on air. Ingenhousz wanted to find out more about the effect plants had on air, so he set up experiments to investigate the gases plants give off under water.

[...]

Ingenhousz performed about 500 experiments on plants and published his findings in 1779 in a book called, *Experiments Upon Vegetables, Discovering Their Great Power of Purifying the Common Air in Sunshine, and of Injuring It in the Shade and at Night*.

[...]

In effect, Ingenhousz had discovered photosynthesis, although he did not name the process, nor did he understand exactly how it worked. He also realized that animals and plants are interdependent: animals (including humans) need oxygen from plants, and plants need the carbon dioxide people and other animals breathe out.

As well as his work on plants and inoculation, he investigated heat conduction and electricity, inventing a machine to create static electricity. He returned to England in 1779 to publish his book and died in 1799 at Bowood in Wiltshire.

Language features and style of the recount text

The Life of Jan Ingenhousz

Jan Ingenhousz was born in Breda, Holland in 1730. He was the son of a leather merchant. Jan was a clever boy and he ended up going to university to study medicine first in Holland, then in Paris and Edinburgh.

He worked as a doctor in his home town until he was 35, when he went to work in a hospital in London. Here he was one of the pioneers of inoculation against the deadly disease, smallpox. In 1768 his work was noticed by the king, George III, who sent him to Vienna to inoculate the Austrian Empress Maria Theresa. Ingenhousz stayed in Austria as court physician until 1779, when he returned to England.

While he was in Austria he studied other subjects, including electricity and plants. His work on plants was very detailed and he did hundreds of experiments to try and find out exactly how they worked.

Ingenhousz had met Priestley when he was working in London and knew all about Priestley's work on air. Ingenhousz wanted to find out more about the effect plants had on air, so he set up experiments to investigate the gases plants give off under water.

[...]

Ingenhousz performed about 500 experiments on plants and published his findings in 1779 in a book called, *Experiments Upon Vegetables, Discovering Their Great Power of Purifying the Common Air in Sunshine, and of Injuring It in the Shade and at Night.*

[...]

In effect, Ingenhousz had discovered photosynthesis, although he did not name the process, nor did he understand exactly how it worked. He also realized that animals and plants are interdependent: animals (including humans) need oxygen from plants, and plants need the carbon dioxide people and other animals breathe out.

As well as his work on plants and inoculation, he investigated heat conduction and electricity, inventing a machine to create static electricity. He returned to England in 1779 to publish his book and died in 1799 at Bowood in Wiltshire.

Annotations (left):

- Written in past tense
- Events placed in time order (general rule)
- Written in first or third person (consistent throughout text)
- Specific focus (e.g. person, place, period)
- Paragraphs divide the recount into sections

Introduction sets the scene – biographical summary

This and following paragraph expand on aspects of Ingenhousz's work

Three dots – called an ellipses – indicate that some text has been cut

Annotations (right):

Focus is on Jan Ingenhousz – specifically his work on plants

Written in third person

Use of words to do with time and time passing

Closing sentence brings recount to a logical conclusion (in this case the death of Ingenhousz)

Content and organization of the recount text

Rather than using one paragraph for each major incident in Ingenhousz's life, this recount introduces Ingenhousz in the first paragraph through a brief retelling of his life from his birth in 1730 to his return to England in 1779. The recount then returns to the period of his life spent in Austria in order to describe Ingenhousz's work on plants relating to photosynthesis. A careful reading of the written recount produces a timeline such as the one below.

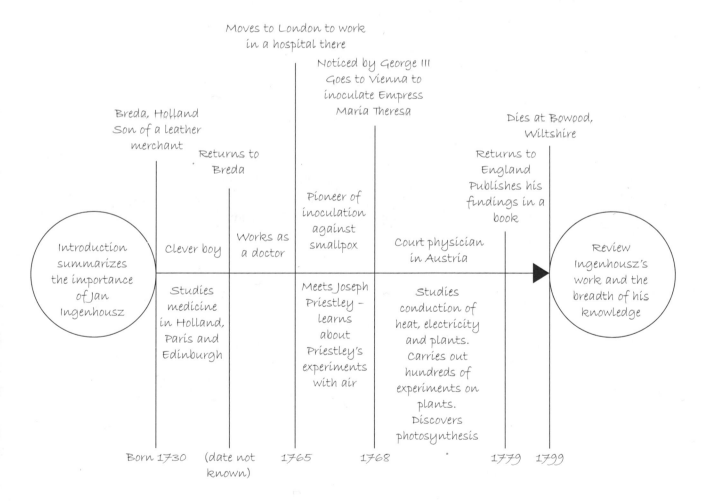

Teaching pupils how to read and write report text

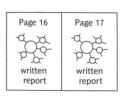

Page 16	Page 17
written report	written report

Reading a report text

Read pp 16–17 of *Interdependence and Adaptation with* the pupils. You will need:

◆ the written report text on p 16 (the reduced text-only version on p 27 of these notes can be enlarged/photocopied/made into an OHT for annotation);

◆ p 26 of these notes enlarged/photocopied/made into an OHT for annotation.

Audience and purpose

> SHARED READING ACTIVITY

Talk about how the intended audience and purpose affects language and layout.

Audience – school pupils, who may know the basic nutritional needs of plants (light, carbon dioxide and water), but be unaware of their need for extra nutrients.

Purpose – to describe the extra nutrients that plants need and why; to explain the consequences if plants do not get these nutrients.

Content and organization

> SHARED WRITING ACTIVITY

Review the content and show pupils how the information in this written report text can be organized as a visual report skeleton (or spidergram) (see p 29 of these notes). Each paragraph becomes one arm of the skeleton with detailed notes around it.

Language features and style

> SHARED READING ACTIVITY

Return to the text and talk about the way language has been used to achieve the effects the author intended (see annotated version on p 28 of these notes). Note useful features for use in pupils' own writing e.g. manipulation of sentence structure for emphasis; use of conditionals (*if . . . then*); use of bullet points for clarity. Add other examples provided by pupils.

> INDEPENDENT WRITING ACTIVITY

Tell the pupils to imagine they are compiling part of a 'frequently asked questions' slot for a gardening magazine. Pupils use the skeleton notes to write questions and answers about the care of plants and the need for nutrients.

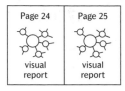

Writing a report text

Use pp 24–25 of *Interdependence and Adaptation* as a basis for the pupils' own report texts.

You will need:

◆ the visual report on pp 24–25;
◆ p 26 of these notes enlarged/photocopied/made into an OHT for annotation.

Content and organization

> **PAIRED READING AND WRITING ACTIVITY**

In pairs, pupils discuss the visual report on pp 24–25 and make skeleton notes (see below). Notes should include the facts listed, details from the pictures and any further information the pupils know. For each feeding type (herbivore, carnivore and omnivore, detritivore, scavenger), consider: What kind of foods does the type eat? How varied is the diet? What physical adaptations does the type have (e.g. teeth, position of eyes, digestive system, colouring, shape)? What are the advantages/disadvantages of each type of diet? Discuss the notes.

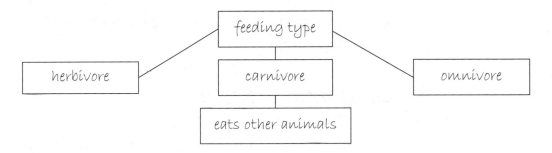

Language features and style

> **SHARED READING ACTIVITY**

Remind pupils of the language features and style of reports (see p 26 of these notes).

Audience and purpose

Discuss the audience for pupils' reports (readers who do not know the differences between the various feeding types) and the purpose (to provide basic information about a typical example of each of the main feeding types).

> **SHARED WRITING ACTIVITY**

Demonstrate writing a short introductory paragraph, for example;
Animals need food to grow, stay healthy and repair their bodies. Different animals are adapted to eat and digest different types of food. Three main types of feeder are herbivores, carnivores and omnivores. Two other important types of feeder are detritivores and scavengers.

Scribe a paragraph about one of the feeding types made up by the pupils. Revisit the questions in content and organization above. Pairs of pupils can provide sentences that answer the questions clearly and precisely. For example;

◆ *A typical herbivore, such as a zebra, eats plants, especially grasses. There is little variety in the diet. As its food is difficult to digest, a zebra has grinding teeth and a specialized digestive system. An advantage of being a herbivore is that you do not have to waste energy hunting your food. A disadvantage is that you spend a lot of time eating because your food is difficult to digest. Also, carnivores hunt you.*

> **INDEPENDENT WRITING ACTIVITY**

Pupils write up the remaining paragraphs independently.

About report text

Audience and purpose

Audience – someone who wants to know about the topic.

Purpose – to describe what something is like.

> Sometimes you may know more about the age or interests of your reader

Content and organization

- **non-chronological** information
- **introductory sentence or paragraph** says what the report is going to be about
- the information is sorted into groups or **categories**
- reports may include short pieces of explanation

> This means it ISN'T written in time order, like a story or recount

> What something looks like, where it is found . . .

Language features

- written in the **present tense**
- usually **general nouns and pronouns** (not particular people or things)
- **factual descriptive words**, not like the descriptions in a story
- words and devices that show **comparison and contrast**
- **third person** writing to make the report **impersonal and formal**
- **technical words and phrases** –which you may need to explain to the reader
- use of **examples** to help the reader understand the technical words

> You would write about dogs in general, not a particular dog

> You would say powerful beams, not beautiful, bright beams

> Expressions like have in common, the same as . . ., on the other hand, however . . .

> Unusual words that go with the topic such as, canine, translucent and wingspan

> Wingspan is the distance between the tips if a bird's outstretched wings

The basic skeleton for making notes is a spidergram

An example of a report text

A Balanced Diet for Your Plants

Is your breakfast cereal 'fortified with vitamins and minerals'? Do you sometimes take food supplements? Do you know anyone expecting a baby? Perhaps she takes supplements to help the development of her growing foetus. We supplement our diet with vitamins and minerals to keep us healthy without giving it too much thought. But it's not only humans that need extra nutrients – plants do too. To ensure your house and garden plants are healthy you need to make sure they have a balanced diet. If they have discoloured patches on their leaves, are not growing, flowering or fruiting well, the chances are they are short of one or more essential nutrients.

Plants get most of the extra nutrients they need from the soil. They need:

◆ iron and copper to make chlorophyll
◆ nitrogen to build their cell walls and make protein
◆ potassium to help control their intake of water and build up their resistance to disease
◆ phosphorus to help photosynthesis and reproduction.

However, sometimes the soil is short of one or more of these vital nutrients. In these cases, the plants will not be able to grow well, and may eventually die.

Language features and style of the report text

- Introduction acts as guide to reader of content
- Present tense throughout
- General nouns and pronouns e.g. **plants, humans, you**
- Factual style
- Non-chronological throughout
- Use of technical words and phrases

A Balanced Diet for Your Plants

Is your breakfast cereal 'fortified with vitamins and minerals'? Do you sometimes take food supplements? Do you know anyone expecting a baby? Perhaps she takes supplements to help the development of her growing foetus. We supplement our diet with vitamins and minerals to keep us healthy without giving it too much thought. But it's not only humans that need extra nutrients – plants do too. To ensure your house and garden plants are healthy you need to make sure they have a balanced diet. If they have discoloured patches on their leaves, are not growing, flowering or fruiting well, the chances are they are short of one or more essential nutrients. . . .

Plants get most of the extra nutrients they need from the soil. They need:

- iron and copper to make chlorophyll
- nitrogen to build their cell walls and make protein
- potassium to help control their intake of water and build up their resistance to disease
- phosphorus to help photosynthesis and reproduction.

However, sometimes the soil is short of one or more of these vital nutrients. In these cases, the plants will not be able to grow well, and may eventually die.

Introduction gets readers' interest by posing questions

Instead of explaining the word, the author makes meaning clear in the next sentence

To aid understanding, author draws on pupils' own experience of dietary supplements and relates it to needs of plants

The dash indicates that what comes next may be surprising

Author assumes some knowledge and understanding of technical terms

Sentence structure and order manipulated for effect

Use of conditional 'If . . . then' (Note the word 'then' is understood, but omitted, after 'fruiting well')

Connective indicates a contrasting statement as a paragraph opening

Bullet points
- add clarity and emphasis
- remove the need for a list using a colon and semi-colons
- are expanded later in the passage

Content and organization of the report text

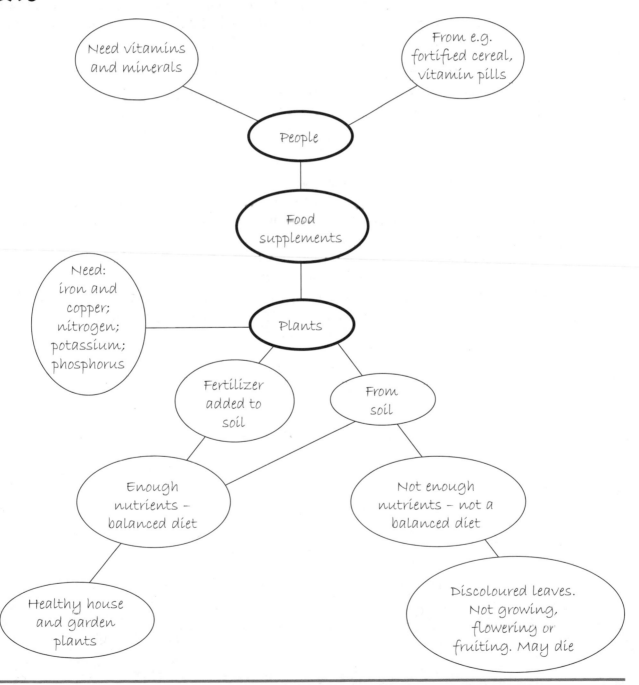

Teaching pupils how to read and write explanation text

Reading an explanation text

Page 26	Page 27
⚬→⚬→⚬	⚬→⚬→⚬
written explanation	written explanation
visual explanation	visual explanation

Read pp 26–27 of *Interdependence and Adaptation with* the pupils. You will need:

◆ the written explanation text on pp 26–27 (the text-only version on p 33 of these notes can be enlarged/photocopied/made into an OHT for annotation);

◆ p 32 of these notes enlarged/photocopied/made into an OHT for annotation.

Audience and purpose

SHARED READING ACTIVITY

Talk about how the intended audience and purpose affects language and layout.

Audience – school pupils who have some knowledge of feeding relationships and the concept of habitats.

Purpose – to explain how food chains can be used to represent feeding relationships within a habitat; to understand that food chains begin with a producer (plant) which gets its energy from the Sun; to introduce the idea that most feeding relationships are more complicated than a simple chain.

Content and organization

SHARED WRITING ACTIVITY

Revise the content. Then demonstrate to pupils how the information in this explanation text is organized by drawing an explanation flowchart (see p 35 of these notes). Point out that information in the flowchart often needs to be supported by diagrams.

Language features and style

SHARED READING ACTIVITY

Return to the text and talk about the way language has been used to achieve the effects the author intended (see annotated version on p 36 of these notes). Note useful features for later use in pupils' own writing e.g. Use of the present tense; general nouns (such as plant, animals); general statement at the beginning of the text to introduce the topic. Add other examples provided by pupils.

INDEPENDENT ACTIVITY

Pupils use their skeleton notes and any diagrams and present a verbal explanation of food chains to their class or group This could also form the basis for a class assembly, involving role-play. (Link to Drama)

Writing an explanation text

Use the section about photosynthesis of *Interdependence and Adaptation* as a basis for the pupils's own written explanation. You will need:

◆ the visual explanation on p 6;

◆ p 32 of these notes enlarged/photocopied/made into an OHT for annotation.

SHARED READING ACTIVITY

Content and organization

Revise the content and organization of explanation text from the previous session (see p 35 of these notes).

INDEPENDENT/ PAIRED ACTIVITY

In pairs, pupils discuss the visual explanation on p 6 and make skeleton notes (see example below) to show how plants obtain their food.

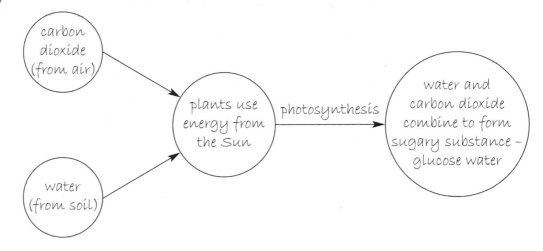

SHARED READING ACTIVITY

Language features and style

Remind pupils of the main language features and style of written explanation text (see p 32 of these notes).

Audience and purpose

Discuss the audience for pupils' written explanations (pupils who can name parts of a plant, but do not know how the plant obtains food) and purpose (to explain clearly how plants obtain their food).

Discuss the visual explanation skeleton as a class. Decide which are successful examples and how they can be used to produce a class skeleton.

SHARED WRITING ACTIVITY

Demonstrate how to create a written explanation by scribing an introductory sentence, using the notes in the visual explanation. For example: *Like animals, plants need food to support life as well as for growth and repair, but unlike animals, green plants obtain food by making their own.*

Continue with the written explanation by scribing suggestions given by the pupils. e.g. *green plants can do this because they are able to use the Sun's energy directly. As a result of their ability to make food, they are known as producers. Food is produced in special cells called chloroplasts that are found mainly in the plant's leaves.*

INDEPENDENT WRITING ACTIVITY

Pupils then use notes in the class visual explanation as a basis for their own written explanation of photosynthesis.

About explanation text

Audience and purpose

Audience – someone who wants to understand the process (how or why it happens).

Purpose – to explain how or why something happens.

Sometimes you may know more about the age or interests of your reader

Content and organization

- **title** often asks a question, or says clearly what the explanation is about
- text often opens with **general statement(s)** to introduce important words or ideas
- the process is then written in a **series of logical steps**, usually in **time order**
- sometimes **picture(s)** or **diagram(s)**

This happens… then this happens… next…

Language features

- **third person** writing to make the explanation **impersonal and formal**
- written in the **present tense**
- usually **general nouns and pronouns** (not particular people or things)
- **factual descriptive words**, not like the descriptions in a story
- **technical words and phrases** – which you may need to explain to the reader
- words and devices that show **sequence**
- words and devices that show **cause and effect**

You would say powerful beams, not beautiful bright beams

You would write about dogs in general, not a particular dog

Unusual words that go with the topic such as, canine, translucent and wingspan

First…, next…, finally

If…, then… This happens because… This means that…

The basic skeleton for making notes is a flowchart

The explanation skeleton can change depending on the sort of process

An example of an explanation text

Chains and pyramids

Energy cannot be destroyed but it can be passed on. That is what happens to the Sun's energy. It is absorbed by plants (producers) during photosynthesis, then passed on to the animals (consumers) that eat them, and on again to the animals that eat those animals – and so on. But at each level some of the energy is needed by the organisms, and some is lost as heat, so there is less and less energy being passed up the chain. This means that fewer animals are supported at each level. This is sometimes shown in a diagram known as a 'pyramid of numbers'.

The feeding relationships between organisms can also be shown in a food chain like these below.

[*see diagram in the Pupils' Book*]

Language features and style of the explanation text

Chains and pyramids

Energy cannot be destroyed but it can be passed on. That is what happens to the Sun's energy. It is absorbed by plants (producers) during photosynthesis, then passed on to the animals (consumers) that eat them, and on again to the animals that eat those animals – and so on. But at each level some of the energy is needed by the organisms, and some is lost as heat, so there is less and less energy being passed up the chain. This means that fewer animals are supported at each level. This is sometimes shown in a diagram known as a 'pyramid of numbers'.

The feeding relationships between organisms can also be shown in a food chain like these below.

[*see diagram in the Pupils' Book*]

- Present tense throughout
- General nouns and pronouns e.g. plants, animals, energy
- Factual, formal style
- Often accompanied by diagrams
- Use of technical words and phrases

Bracketed words remind reader of feeding relationship between plants and animals

Use of a causal connective (indicates cause and effect)

Reference to explanatory diagram

Title is a guideline to what is being explained

Text begins with a general statement to introduce the topic

Use of technical terms

Text refers directly to the relevant diagram – a diagram adds to our understanding of the text. It is not just a captioned illustration

Importance of accompanying diagrams – it is difficult to explain an abstract idea, such as a food chain, using words alone

If you are using this text with other year groups then also highlight these features:

Y3/P4　◆　Formal style – requires very little use of adjectives.

Y4/P5　◆　Use of degrees of intensity e.g. **less** and **less** energy; **more** complicated.

Y5/P6　◆　Use of passive voice e.g It is **trapped**; animals **are supported**.

Content and organization of the explanation text

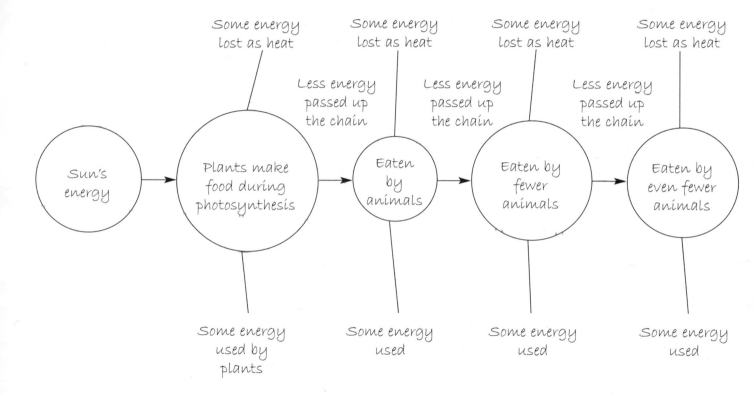

Teaching pupils how to read and write discussion text

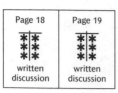

Page 18	Page 19
written discussion	written discussion

Reading a discussion text

Read pp 18–19 of *Interdependence and Adaptation* with the pupils. You will need:

- the written discussion text on pp 18–19 (the reduced text-only version on p 39 of these notes can be enlarged/photocopied/made into an OHT for annotation);
- p 38 of these notes enlarged/photocopied/made into an OHT for annotation.

Audience and purpose

SHARED READING ACTIVITY

Talk about how the intended audience and purpose affects language and layout.

Audience – anyone with only a vague idea of what genetic modification means.

Purpose – to explain what genetic modification is, and to put forward a balanced view of the advantages and disadvantages of using GM foods.

Content and organization

SHARED WRITING ACTIVITY

Revise the content and show pupils how the information in this discussion text can be organized, using notes in a for-and-against grid (see p 41 of these notes).

Language features and style

SHARED READING ACTIVITY

Return to the text and talk about the way language has been used to achieve the effects the author intended (see annotated version on p 40 of these notes). Note useful features for use in pupils' own writing e.g. use of 'balancing connectives' (*however, on the other hand),* use of present tense and impersonal style. Add other examples provided by pupils. Look at the full text in the pupils' books to find further examples and additional support for the arguments.

INDEPENDENT ACTIVITY

After they have studied the skeleton carefully, ask pupils, in pairs, to play a game called 'good news/bad news'. One pupil has to state an advantage of GM (the good news) and the other has to respond with a corresponding disadvantage (the bad news), or vice versa with the bad news being given first e.g. *"The good news is that farmers can grow better crops more quickly." "The bad news is that genes from these crops could affect other plants."*

After the pupils have done this for a while, bring them back together to play the game as a class, ensuring that the 'good news' and 'bad news' are balanced arguments for and against genetic modification

Page 20	Page 21
written report	visual discussion

Writing a discussion text

Use pp 20–21 of *Interdependence and Adaptation* as a basis for the pupils's own written discussion. You will need:

◆ the visual discussion on p 21;

◆ p 38 of these notes enlarged/photocopied/made into an OHT for annotation.

Content and organization

Revise the content and organization of discussion text from the previous session (see p 41 of these notes).

PAIRED READING AND WRITING ACTIVITY

In pairs, pupils study the visual discussion on p 21. They should consider the following points. Which advantages/disadvantages 'cancel each other out'? For example, the advantage that *fertilizers put nutrients back into the soil once crop has been harvested* is balanced by the counter argument *(chemical fertilizers) do not replace organic matter needed to keep soil structure healthy*.

Are there any advantages/disadvantages that have no counter argument? How important is this?

SHARED WRITING AND READING ACTIVITY

Ask some of the pupils to 'hot seat', playing the roles of large-scale users of fertilizer, and opponents of fertilizer use, answering questions from the rest of the class. Use these activities to compile a class for-and-against grid (for an example, see p 41 of these notes).

Language features and style

Remind pupils of the language features and style of discussion texts (see p 38 of these notes).

Audience and purpose

Discuss the audience for pupils' discussion texts (readers who know about the purpose of fertilizers and who want to be given facts so they can make up their own mind on a topic) and the purpose (to write a newspaper article or multi-media presentation giving a balanced report on the large-scale use of fertilizers).

SHARED WRITING ACTIVITY

Look again at the opening and closing paragraphs of the newspaper article on pp 18–19 in the pupils' book. Then model an introductory sentence e.g. *The large-scale use of chemical fertilizers can greatly increase food production and holds out the hope that, in the developing world, famine can be prevented.* Scribe the rest of the introductory paragraph using suggestions from the class.

INDEPENDENT WRITING ACTIVITY

Pupils write up the remaining paragraphs independently, referring to the class for-and-against grid for guidance about the structure of the article and paragraph divisions. The discussion should summarize the conflicting views, analyse the arguments for and against GM fairly, and distinguish clearly between balanced reporting and personal opinion.

ICT can be used to plan, revise and edit the writing, and to develop the layout and presentation into the style of a newspaper.

The work could also be developed as a multimedia presentation. (Links to ICT QCA Unit 6a)

About discussion text

Audience and purpose

Audience – someone who wants to know both sides of the argument, but may not know much about the subject.

Purpose – to present arguments and information from different viewpoints.

> Sometimes you may know more about the age or interests of your reader.

Content and organization

- usually starts with a sentence or paragraph **introducing the subject** under discussion and **defining important terms**
- the argument is then split into a number of **main points for and against**
- the arguments for and against are supported by **evidence and examples**
- **concluding sentence or paragraph** sums up the main points, for and against (and sometimes expresses the author's own opinions)

> Introduce important words or ideas the reader needs to know.

> You can either give all the arguments for, then all the arguments against, or all arguments for and against each point – one by one.

> The examples could:
> – agree with the point
> – back up the evidence
> – add further information to explain it.

Language features

- written in the **third person** to make the discussion **impersonal and formal**
- usually in the **present tense**
- usually **generalized nouns** (except in specific examples)
- words and devices showing **cause and effect**, used to **argue** the case
- words and devices that signal a **move from one side of the argument** to the other
- words and devices which suggest '**possibility**' rather than certainty

> Give the people on each side names, such as Supporters claim..., Critics reply...

> However..., On the other hand...

> Environ-mentalists, developers, scientists...

> Therefore..., Consequently..., This means that...

> Perhaps..., probably..., might..., could be...

The basic skeleton for making notes is a for-and-against grid

An example of a discussion text

The great GM debate

GM scientists argue that they are merely doing what farmers have been doing for generations – cross-breeding plants to make better crops, but doing it more quickly and efficiently. On the other hand, environmentalists are concerned that when GM crops are grown in the wild, the modified genes might 'escape' into the environment. This could result, for example, in wild species becoming resistant to pesticide – creating 'superweeds' which cannot be destroyed by standard herbicides.

Another point in favour of GM foods is that they can have positive health benefits, for instance by increasing the storage-life of foods. Crops can also be made more nutritious. . . .

However, scientists point to many more advantages of GM crops. They can produce higher yields, which could benefit both farmers and poorer nations that need efficient food production. They also require less fertilizer, which will reduce environmental pollution. They can even be engineered to tolerate extreme environments – such as heat, cold or salty conditions.

[. . .]

Those in favour of GM are quick to point out that it is not just about yield: it can make crops more resistant to pests and diseases, so that fewer pesticides are used. This means less pesticide getting into the environment – and into our bodies via food. On the other hand, GM critics argue that pest-resistant crops actually produce their own insecticide, and that this will also get into the human food chain.

[. . .]

The final argument in favour of genetic modification. is that it allows scientists to transfer certain characteristics from one plant to another. It can even give plants characteristics taken from other organisms. . . .

There is little doubt that GM has the potential to help us feed the world efficiently, with fewer environmental consequences, if it is developed responsibly. Nevertheless, its critics argue that long-term effects of genetic modification are not sufficiently understood, and that GM organisms need long and careful testing before being released into the environment.

Language features and style of the discussion text
The great GM debate

- Present tense used for main body of the text (see below)
- Balanced presentation
- Factual, formal style
- Non-chronological

Indicates deduction or speculation

Formal writing – use of passive voice

By opening with two initial letters that arouse controversy, the writer attracts the reader's attention straight away

Connectives indicate that writing is not biased. Each argument is balanced by a counter-argument

Shows clearly who supports the argument

Three dots (called an ellipsis) indicate some text has been cut

In concluding paragraph, writer expresses own opinion, but it is still a balanced one

GM scientists argue that they are merely doing what farmers have been doing for generations – cross-breeding plants to make better crops, but doing it more quickly and efficiently. On the other hand, environmentalists are concerned that when GM crops are grown in the wild, the modified genes might 'escape' into the environment. This could result, for example, in wild species becoming resistant to pesticide – creating 'superweeds' which cannot be destroyed by standard herbicides.

Another point in favour of GM foods is that they can have positive health benefits, for instance by increasing the storage-life of foods. Crops can also be made more nutritious. . . .

However, scientists point to many more advantages of GM crops. They can produce higher yields, which could benefit both farmers and poorer nations that need efficient food production. They also require less fertilizer, which will reduce environmental pollution. They can even be engineered to tolerate extreme environments – such as heat, cold or salty conditions.

[. . .]

Those in favour of GM are quick to point out that it is not just about yield: it can make crops more resistant to pests and diseases, so that fewer pesticides are used. This means less pesticide getting into the environment – and into our bodies via food. On the other hand, GM critics argue that pest-resistant crops actually produce their own insecticide, and that this will also get into the human food chain.

[. . .]

The final argument in favour of genetic modification. is that it allows scientists to transfer certain characteristics from one plant to another. It can even give plants characteristics taken from other organisms. . . .

There is little doubt that GM has the potential to help us feed the world efficiently, with fewer environmental consequences, if it is developed responsibly. Nevertheless, its critics argue that long-term effects of genetic modification are not sufficiently understood, and that GM organisms need long and careful testing before being released into the environment.

If you are using this text with other year groups then also highlight these features:

Y3/P4
- ◆ Length of sentences. There are very few short, simple sentences. Look at variety of ways of creating complex sentences. Ask questions about this and the range of vocabulary. What kind of audience is the article intended for?

Y4/P5
- ◆ The writer uses language to distinguish between fact and opinion: e.g. That GM crops 'produce higher yields' is a *fact*; 'on the grounds that it is too dangerous' is *opinion*.

Y6/P7
- ◆ Wide-ranging use of punctuation, helping the reader to understand the meaning of long, complex sentences and to make such sentences more interesting. e.g. as well as commas, hyphens and semicolons are used. Informal words (**escape**, **superweeds**, and *golden rice*) used instead of technical words are put between quotation marks to avoid confusion.

Content and organization of the discussion text

What are the advantages and disadvantages of genetic modification?

Advantages	Disadvantages
◆ Better crops produced more quickly than traditional cross-breeding.	◆ If genes 'escape' – weeds that resist herbicides.
◆ Health benefits (longer storage life; can be more nutritious).	◆ If GM genes go to bacteria – resistant bacteria.
◆ Higher yields.	◆ If dangerous – hungry people won't use food.
◆ Tolerate extreme environments – food in more areas e.g. drought areas.	◆ Insecticide from GM crops in food chain.
◆ Fewer pesticides in environment and humans. Less fertilizer so less pollution in environment.	◆ Insects may become resistant to insecticides.
	◆ GM insecticide harms both 'good' and 'bad' insects.

Teaching pupils how to read and write reference texts

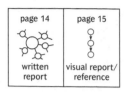

page 14	page 15
written report	visual report/ reference

Reading reference texts

Read p 15 of *Interdependence and Adaptation with* the pupils. You will need:

- the reference text about specialist aerial roots in the chart on p 15 (the text-only version on p 45 of these notes can be enlarged/photocopied/made into an OHT for annotation);
- p 44 of these notes enlarged/photocopied/made into an OHT for annotation.

> SHARED READING ACTIVITY

Audience and purpose

Audience – pupils who are studying the adaptation of roots.

Purpose – to obtain information about specialized roots that grow above ground.

Content and organization

> SHARED WRITING ACTIVITY

The information in the text is very specific. It relates only to roots that are adapted to grow above ground and which perform particular functions (report/explanation text). The information is divided into entries, each with a heading. The headings are arranged *alphabetically*.

The text in the chart is written in note form as it is also an example of visual report text. Pupils could turn the notes into full sentences and paragraphs to create an example of written reference text.

Not all reference texts are arranged in this way, and their content varies widely (see *About reference text* on p 44*).* However, all are structured and organized in some way that helps the reader to find the information they are looking for.

Brainstorm other types of reference texts organized in a similar way to the information on roots i.e each entry having a heading followed by the main text. Discuss the different suggestions and the purpose of each.

Language features and style

> SHARED READING ACTIVITY

Return to the text and talk about the language features (see annotated version on p 46 of these notes). In particular look at the features which make each entry clear, and the relevant information easy to find *e.g. alphabetical ordering; separation of headings from main part of entry; use of bullet points*

> INDEPENDENT ACTIVITY

- Using the results of the brainstorming activity, and any other reference texts you wish to add, have a quiz about sources. Ask the pupils where they would find a certain piece of information e.g. *Where would you find: the telephone number of a garden centre; what 'parasitic' means; how a buttress supports a castle wall; the name of a plant that has 'breathing roots'?*
- Using class dictionaries, a thesaurus, pupils' books, Internet, encyclopaedias, CD-ROMs etc. have a practical quiz. Pupils find the answers to questions, stating where they found the answers. Some information can be found in more than one place (e.g. a dictionary and a glossary might both have an entry on nutrients; an encyclopædia and the Internet might both reveal information about buttresses). Some questions may need extra research. Information about aerial roots, for example, may not be available in a general encyclopaedia.

Writing a reference text

Use the glossary on p 46 of *Interdependence and Adaptation* as a basis for pupils' own written reference text. You will need:

◆ the glossary on p 46;
◆ p 44 of these notes enlarged/photocopied/made into an OHT for annotation.

Content and organization

**PAIRED SHARED
READING AND
WRITING
ACTIVITY**

In pairs, pupils study the glossary on p 46. They should consider features such as:

◆ the purpose of the glossary;
◆ the length and style of the entries;
◆ alphabetic order.

Discuss the pupils' ideas. Then ask the pairs, using pupils' book and their own experience of the unit, to jot down words not in the glossary, but which they think it would be useful to include.

Discuss these together and decide as a class which to include in an extended glossary.

**SHARED
WRITING
ACTIVITY**

Language features and style

Remind pupils of the language features and style of reference texts (see p 44 of these notes).

Audience and purpose

Discuss the audience for pupils' reference texts (Year 6 pupils who want to know the meaning of terms and words used in *Interdependence and Adaptation)* and the purpose (to provide a short, clear explanation of these terms and words).

**SHARED
WRITING
ACTIVITY**

Demonstrate by writing a definition of one of the terms or words requested by the class e.g. *mangrove – marshy area along the cost in tropical climes.* Take suggestions for other glossary terms from the class. Pupils amend them, if necessary, to make them as clear and concise as possible.

**INDEPENDENT/
SHARED WRITING
ACTIVITY**

Pupils write their own definitions for remaining glossary terms, taking one, two or more each depending on attainment levels. Share them as a class, amending them if necessary. Then ask pupils working alone to organize the entries alphabetically, incorporating new entries into the existing glossary.

About reference text

Audience and purpose

Audience – a general reader who wants to find information about a particular subject in a text covering a wide range of similar subjects.

Purpose – to present information on different subjects in a way which is easy to retrieve.

Text structure

- **short pieces of information** (called 'entries') about large numbers of different things, events, people, etc.
- clear system of **organization of entries**, e.g. *alphabetical order (dictionaries), topics (thematic encyclopaedias, web sites), dates (historical encyclopaedias)*
- entries may be **different text types** (report, recount, etc.)
- does not always have an introduction
- may not need contents or index pages

Language features and style

- **impersonal**, usually **formal** style
- clear, **concise** language, e.g. *adjectives and adverbs chosen for clarity rather than vividness or effect*
- entries may be in **present tense** or **past tense**
- may use **generalized nouns, or particular** people or things
- **logical connectives**, e.g. *however, because*

The basic skeleton for making notes is a flowchart

An example of a reference text

Specialist aerial roots

Some plants have special roots which grow above ground. As well as taking in water and nutrients, they can help to hold up the tree, or obtain oxygen.

Root type	Description
Breathing roots	◆ found on plants in swampy areas where there is not enough oxygen in the water for the root to survive ◆ grow above the water or mud and are almost hollow, allowing air to pass directly into the root tissue
Buttress roots	◆ board-like growths on the base of large trees in shallow soils ◆ help support the tree, like the buttresses that support castle walls
Clasping roots	◆ found on plants such as poison ivy to attach the plant to other trees – also absorb nutrients from surface of tree ◆ sometimes grow to the ground
Prop roots	◆ grow above ground from base of stem or branch and act as support
Stilt roots	◆ support trees in swampy areas such as mangroves during changes in mud level (Mangroves are marshy areas along the coast in tropical climes.)

Language features and style of written reference

Specialist aerial roots

Heading and short introduction indicate area of knowledge covered by entries

Some plants have special roots which grow above ground. As well as taking in water and nutrients, they can help to hold up the tree, or obtain oxygen.

Clear separation of heading from entry

Entries organized into alphabetical order

Root type	Description
Breathing roots	- found on plants in swampy areas where there is not enough oxygen in the water for the root to survive - grow above the water or mud and are almost hollow, allowing air to pass directly into the root tissue
Buttress roots	- board-like growths on the base of large trees in shallow soils - help support the tree, like the buttresses that support castle walls
Clasping roots	- found on plants such as poison ivy to attach the plant to other trees – also absorb nutrients from surface of tree - sometimes grow to the ground
Prop roots	- grow above ground from base of stem or branch and act as support
Stilt roots	- support trees in swampy areas such as mangroves during changes in mud level (Mangroves are marshy areas along the coast in tropical climes.)

Adjectives used for clarity, not to add vivid descriptions

Use of examples to support explanation

Use of passive voice

General terms to qualify a statement (lets the reader know that this is not the only support for the tree, and that clasping roots do not always grow to the ground)

Use of bullet points for clarity

If you are using the text on pp 45–46 with another year group then also highlight these features

Y3/P4 ◆ The way the entries are organized is always maintained. For example, 'Buttress roots' comes after 'Breathing roots' maintaining the alphabetical order.

Y4/P5 ◆ Separation of headings from the main part of the entry means that even without the introduction it would be possible to say what the entries were about by scanning the headings.

Y5/P6 ◆ Use of bullet points makes skimming through the text easier by separating where specialist roots are found from their purpose.

Finding and evaluating information quickly

Key words

Pupils in Y6/P7 need to be confident about using key words both to search for information and in their own writing. They play a major role in ICT (when setting up search parameters for CD-ROMs and the Internet, for example) and are a valuable aid when searching for and evaluating information quickly from a written text.

Key words add clarity to writing and are often looked for by markers of national tests and examinations in written answers.

Scanning and skimming

Useful ways of searching for key words and for assessing the value of information are:

Scanning – involves a systemic search for e.g. a key word, without the need to read other information.

Skimming – involves rapid reading without paying close attention to detail to get an impression of what is written. Skimming allows the reader to make a quick judgement about the content and value of information.

A good way of scanning or skimming is to let the eye travel down the centre of a page or column (using a finger as a guide if necessary) either searching for a word or phrase, or rapidly reading what is written on either side.

Suggested activities

Using the index on p 48 of the pupils' book find a suitable word e.g. *photosynthesis*, or *pollinate*. Demonstrate checking the page number and scanning the relevant page to find the key word. Choose another word and let pupils find that word.

Let the class practise the skill on their own with a few more words.

Look together at p 8 of the pupils' book containing visual instructions for an experiment. Suggest a key word or phrase to describe the first instruction (*e.g. plants; two plants; leafy plants*). Ask for suggestions for a key word or phrase to summarize the second instruction (*e.g. light and dark*).

◆ Pupils think of key words or phrases for the final three instructions. Discuss these together, making sure pupils understand why the key words they have chosen are suitable/unsuitable.

◆ The pupils could then use the key words and an instruction skeleton (simple flow diagram) to write their own instructions for the experiment ensuring that the key words they have chosen are incorporated.

SHARED READING ACTIVITY

INDEPENDENT READING ACTIVITY

SHARED READING ACTIVITY

INDIVIDUAL/ PAIRED ACTIVITY

INDEPENDENT WRITING ACTIVITY

Page	Contents	Text type	National Literacy Strategy Objectives	QCA Science Objectives Unit 6a Interdependence and Adaptation
				Pupils should learn
2	**Concept map and Contents**	Reference		
4	**The power source**	Visual report	T1 TL 12,13 T1 SL 2, 3, 4, 5, 6	• that green plants need light in order to grow well
6	**Staying alive**	Visual explanation	T3 TL 15, 19 T3 SL 1, 3, 4	• that green plants need light in order to grow well • to make careful observations of plant growth and to explain these using simple scientific knowledge and understanding
8	**In the dark … and into the light**	Written + visual instructions	T3 TL 16, 19, 20, 22 T3 SL 1, 3 T3 WL 1, 2, 3	• that green plants make new plant material using air, water in the presence of light
10	**The power of plants**	Written recount	T1 TL 12, 14 T1 SL 4, 5, 6 T1 WL 6, 7, 10 T3 TL 21	• that for this to take place the green plant requires leaves
12	**Down to earth**	Written + visual report	T1 TL 12, 13, 17 T1 SL 2, 3, 4, 5, 6 T1 WL 2, 3, 6	• that different plants grow in different soil conditions • that water and nutrients are taken in through the root • that roots anchor the plant in the soil
14	**Getting to the bottom of things**	Written + visual report Visual report/reference	T3 TL 15, 17, 18, 19, 20, 22 T3 SL 1 T3 WL 1, 2, 3	• to make careful, relevant observations of soils • to draw conclusions from observations and to explain these using scientific knowledge and understanding
16	**Food supplements**	Written report	T3 TL 16, 19 T3 SL 1, 3, 4	
18	**The great GM debate (journalistic)**	Written discussion	T1 TL 11, 12,13 T1 SL 2, 3, 4, 5, 6 T1 WL 5, 6 T2 15, 16, 18, 19 T2 SL 1, 3, 4, 5 T2 WL 5, 8	
20	**The effects of fertilizers**	Written report Visual discussion	T2 TL 15, 16, 18, 19 T2 SL 1, 3, 5 T2 WL 2, 3, 8	• that fertilizers are often added to soils to provide plants with the nutrients they need
22	**Fertilizer – bane or boon**	Written + visual persuasion	T2 TL 15, 18 T2 SL 5 T2 WL 8 T3 TL 17, 18, 19 T3 SL 1, 2 T3 WL 2, 3	
24	**You are what you eat**	Visual report	T1 TL 12, 13, 17 T1 SL 2, 3, 4, 5, 6 T1 WL 2, 3, 6 T3 TL 15, 16, 19 T3 SL 1, 3, 4	• that food chains can be used to represent feeding relationships in a habitat • that food chains begin with a plant (the producer)
26	**Chains and pyramids**	Written + visual explanation	TI TL 14, 15, 16, 17 T1 TL 1, 4, 5, 6 T1 WL 7, 8, 9, 10 T3 TL 15, 16, 20, 21, 22 T3 SL 1, 3, 4 T3 WL 2, 3	
28	**The cycle of life**	Written + visual explanation	T2 15, 16, 18, 19 T2 SL 1, 3, 4, 5 T2 WL 5, 8 T3 TL 15, 19 T3 SL 1, 3, 4	• that animals and plants in a local habitat are interdependent
30	**Identification keys**	Written + visual explanation	T3 TL 15, 19 T3 SL 1, 3, 4	• to use keys to identify animals and plants in a local habitat
32	**Adapt or die**	Visual report Visual explanation	T3 TL 15, 19 T3 SL 1, 3, 4 T3 TL 16, 20, 21, 22 T3 SL 1, 3, 4 T3 WL 2, 3	• how animals and plants in a local habitat are suited to their environment
34	**Evolution**	Written report	T1 TL 12, 15 T1 SL 4, 5 T1 WL 6, 7 T3 TL 17, 18	
36	**The life of Charles Darwin**	Visual recount	T1 TL 11, 14, 15, 16 T1 SL 1, 2, 3, 4, 5, 6 T1 WL 2, 3, 6	
38	**Finding your niche**	Written report	T3 TL 19, 22 T3 SL 1, 3, 4	• that different animals and plants are found in different habitats • how animals and plants in a second habitat are suited to their environment
40	**Endangered species**	Written report	T3 TL 16, 17, 18, 20, 21, 22 T3 SL 1, 3, 4	
42	**Adapting to your environment**	Written + visual report	T1 TL 12, 13, 17 T1 SL 2, 3, 4, 5, 6 T1 WL 2, 3, 6	
44	**Living in Madagascar**	Written + visual report	T1 TL 12, 13, 17 T1 SL 2, 3, 4, 5, 6 T1 WL 2, 3, 6 T3 TL 15, 16, 20, 21	
46	**Glossary**	Reference	T3 TL 17, 18, 19, 22 T3 WL 2, 3	
47	**Bibliography**	Reference		
48	**Index**	Reference		